Animal Faces
In the Water

Hannah Kate Sackett

Illustrated by Martin Camm

 **Children's Publishing**

Columbus, Ohio

 **Children's Publishing**

This edition published in the United States in 2003 by
McGraw-Hill Children's Publishing,
A Division of The McGraw-Hill Companies
8787 Orion Place
Columbus, Ohio 43240-4027

www.MHkids.com

Library of Congress Cataloging-in-Publication Data is on file with the publisher.

Created and produced by Firecrest Books Ltd
in association with Martin Camm and Hannah Sackett.

Art and Editorial Direction by Peter Sackett
Edited by Norman Barrett
Edited in the U.S. by Joanna Schmalz and Catherine Stewart
U.S. Production by Tracy Paulus and Nathan Hemmelgarn
Designed by Phil Jacobs
Color Separation by SC International Pte Ltd, Singapore

Printed in Dubai.

ISBN 1-57768-417-6

1 2 3 4 5 6 7 8 9 10 FBL 06 05 04 03 02

Contents

Otter

The morning sun shines on a line of ripples breaking the smooth surface of the river. First a nose, and then a small, flat head emerges from the water. The otter is hunting for food. While the otter is swimming, it closes its nostrils and ears tightly so that no water can get in. The otter moves along onto the seashore. Its keen sense of smell helps it to seek out food around the rock pools.

Shore Crab

When the tide is out, many plants and animals hide in the cracks and crevices of the rock pools. Among them is the shore crab. The crab's strange eyes stick out from its shell on stalks. This feature allows the shore crab to hide its body under rocks, among seaweed, and beneath the sand while still looking for small creatures to feed on.

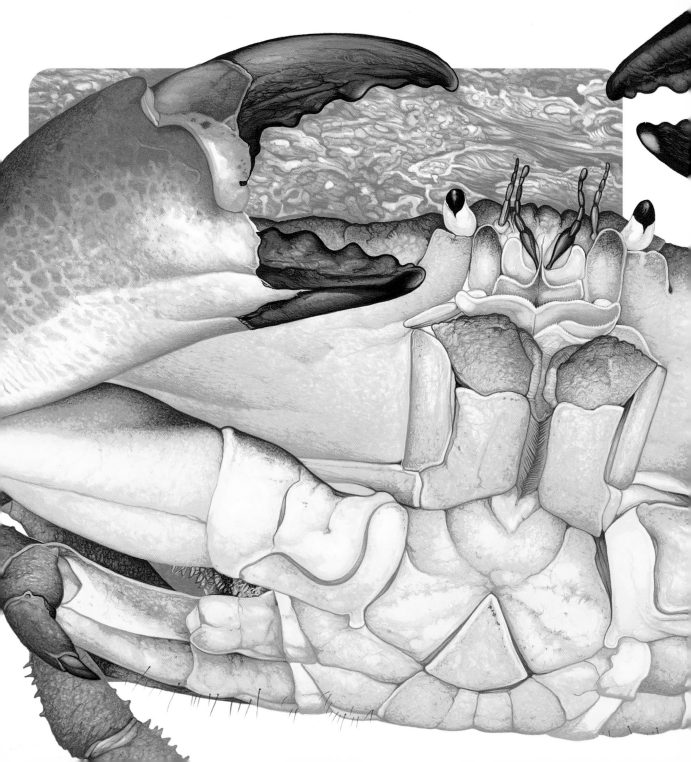

Puffin

Farther out from the beach, where the rugged coast juts into the sea, a crowd of puffins are on the lookout for fish. These birds are skilled hunters, and dive underwater to catch their prey. Their gently hooked beaks can hold as many as twelve fish at a time. The bright beaks of the adult puffins are most colorful in the summertime when they are red, yellow, and blue. The puffins share the rocky shoreline with a group of seals.

Common Seal

Seals can be found basking in the sunshine with their faces turned up toward the sun. These smooth-headed creatures, which are also called harbor seals, have their ears inside their heads. In spite of this, they have good hearing. A young pup can find its mother in a crowd of hundreds by listening out for her special call. Seals spend most of their time out at sea, where their sensitive hearing is even better.

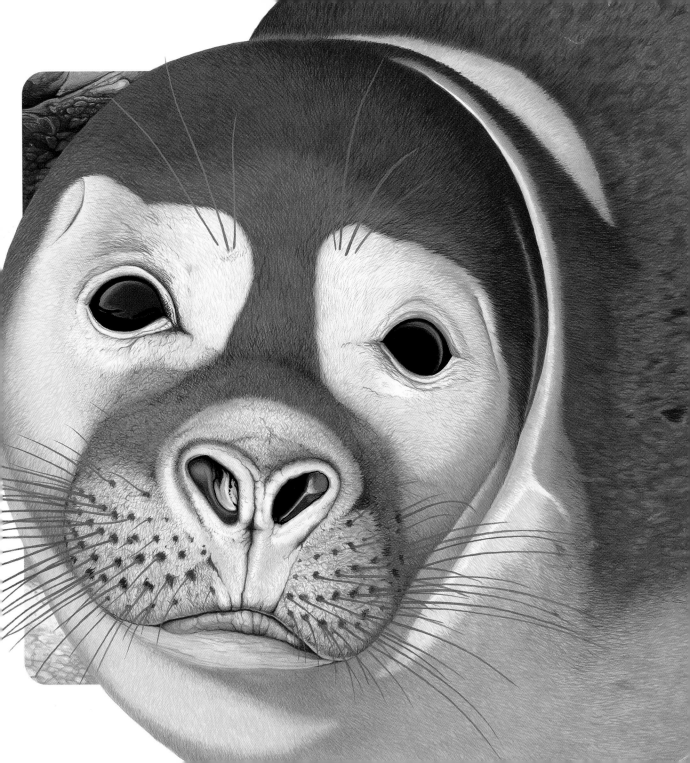

Fulmar

Out at sea, birds are bobbing up and down on the water's surface. Among them are many fulmars. These birds can be seen bowing their heads and cackling as they pass by other birds and warn off enemies. If they or their chicks are in danger, they surprise the attacker by spitting out a stream of foul-smelling liquid. Fulmars spend many hours each day floating on the sea while looking for small fish and scraps of food.

White-tailed Sea Eagle

Unlike the fulmar, the white-tailed sea eagle uses an active approach to catching food. As it soars over the waves, its sharp eyes note the movements of fish beneath the surface. Once in a while, it will drop in a steep dive toward the water to snatch its prey from the shallows. Its powerful, hooked beak makes it a fierce hunter of seabirds and fish.

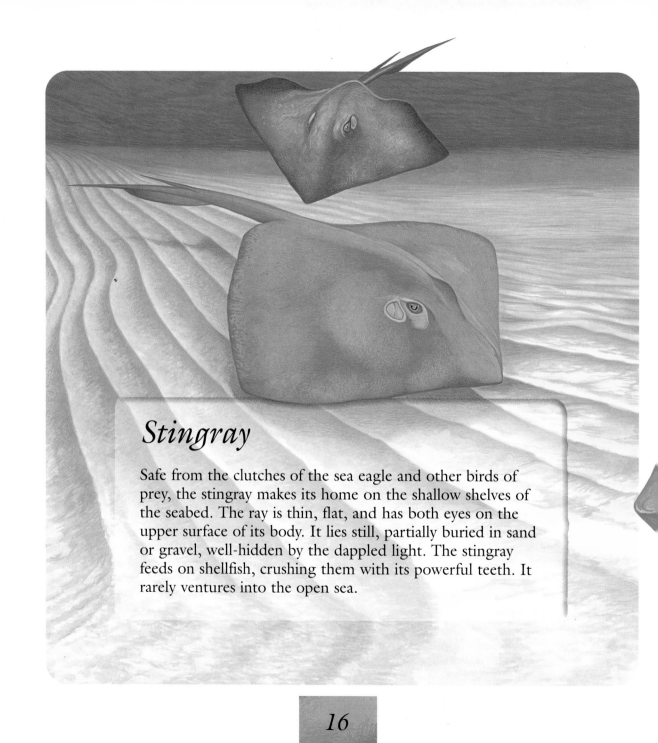

Stingray

Safe from the clutches of the sea eagle and other birds of prey, the stingray makes its home on the shallow shelves of the seabed. The ray is thin, flat, and has both eyes on the upper surface of its body. It lies still, partially buried in sand or gravel, well-hidden by the dappled light. The stingray feeds on shellfish, crushing them with its powerful teeth. It rarely ventures into the open sea.

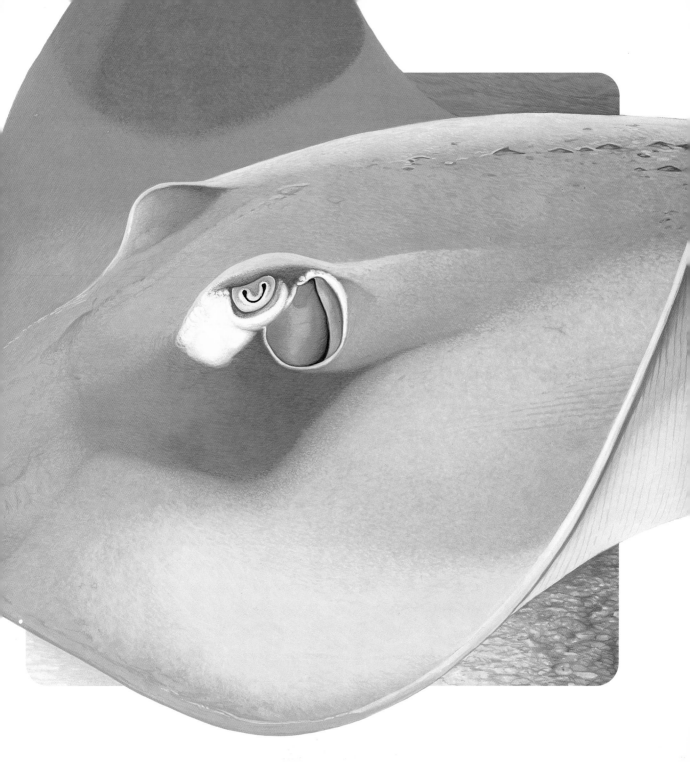

Common Dolphin

Dolphins spend most of their lives in the open sea. The common dolphin is known for its long beak, unusual markings, and amazing ability to find its way around in the darkest water. Instead of using its eyes to see, the dolphin makes clicking noises in its forehead that spread throughout the water. The noises echo off of objects surrounding the dolphin, allowing it to sense the size and location of obstacles as well as the fish and squid that it eats.

Cod

The cod has its own method of finding food in the darkened depths of the sea. On its chin is something called a sensory barbel. Shaped like a hook, the barbel is equipped with taste organs that help the cod hunt down its prey in the half-darkness.

Deep-sea Shrimp

The deep-sea shrimp lets off an eerie glow as it moves through the water. Chemicals in the shrimp's body produce these yellow-green flashing lights that are used to attract mates. Some deep-water shrimp squirt chemicals from under their eyes when threatened. These chemicals light up the water around them, possibly warning off attackers.

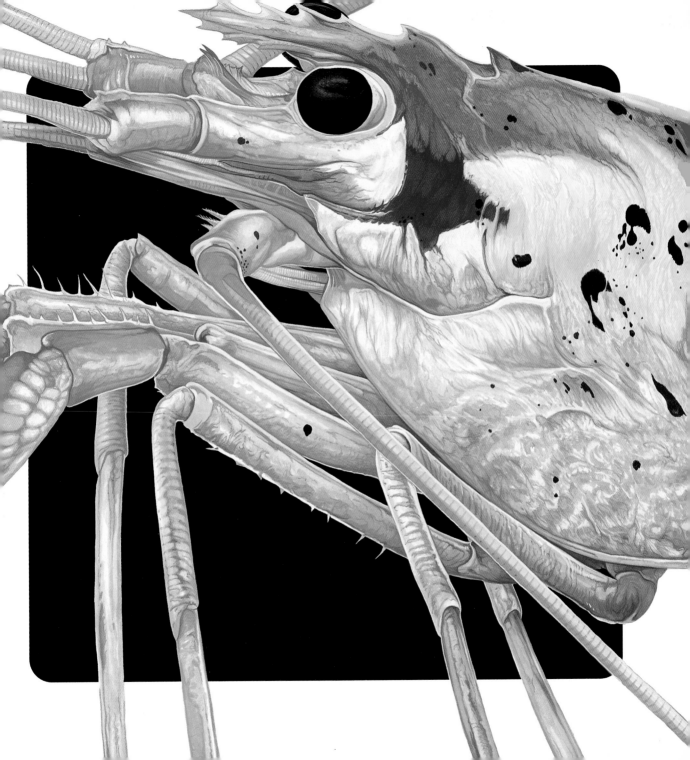

Anglerfish

One of the most interesting fish to use underwater lighting is the slow-moving anglerfish. Just like a fisherman, it uses a tempting bait to attract passing fish. The bait is a glowing ball attached to the end of a flexible spine, which extends out from above its mouth. The light from the ball illuminates approaching fish. As the anglerfish opens its wide mouth, the prey is sucked into its gaping jaws.

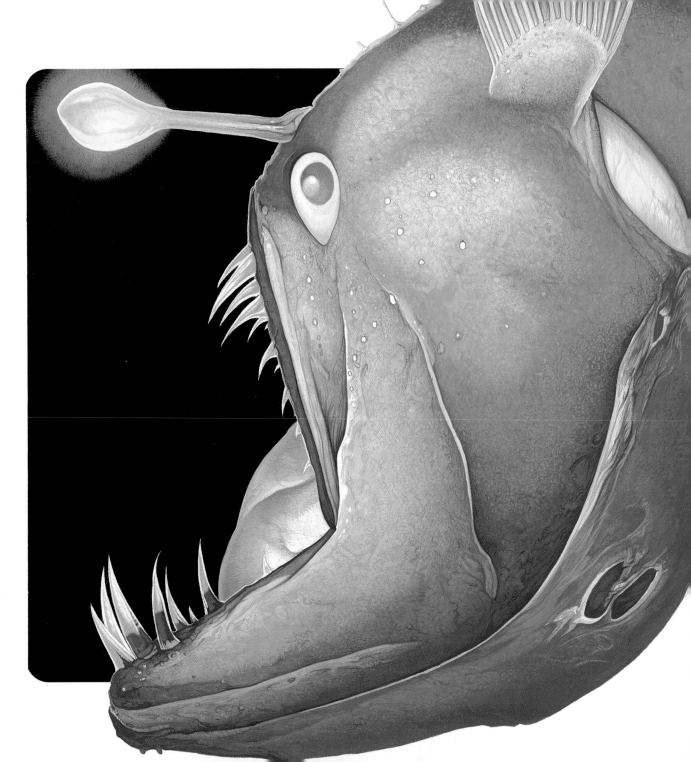

Fang-tooth Fish

Another threat to small sea creatures is the fang-tooth fish. It gets its name from the long, sharp fangs growing from its upper and lower jaws. Like some snakes, these fish can unhinge their jaws in order to catch and swallow large prey. Fang-tooth fish are one of the unusual animals able to survive in the darkness of the deep.

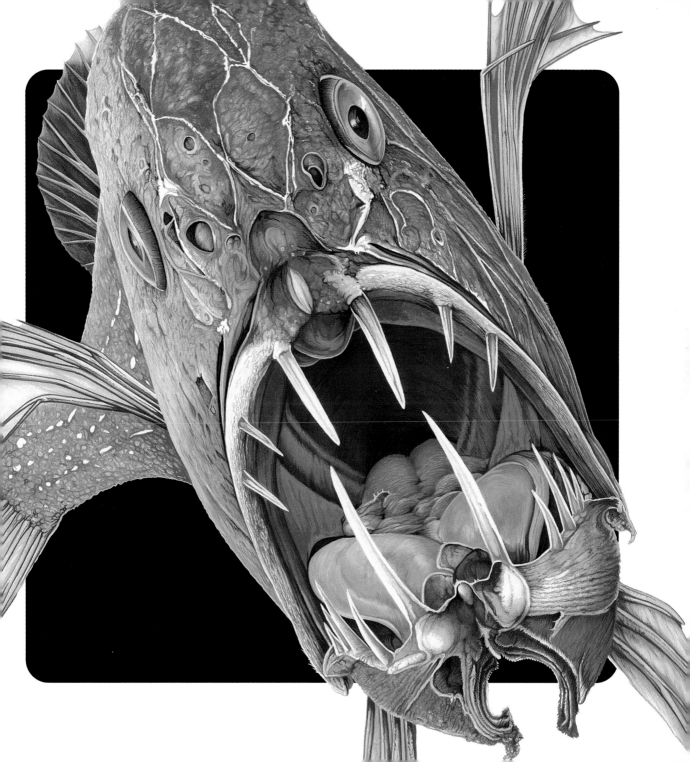

Sperm Whale

The enormous sperm whale dives to unknown depths to hunt for giant squid. With a head measuring up to 20 feet in length, it is no wonder the sperm whale has one of the largest faces in all of the animal kingdom. This impressive sea mammal breathes through a single nostril called a blowhole, located on the top of its head. When a sperm whale comes up from the depths of the ocean to the surface, it breathes out, sending a spectacular cloud of spray into the air.

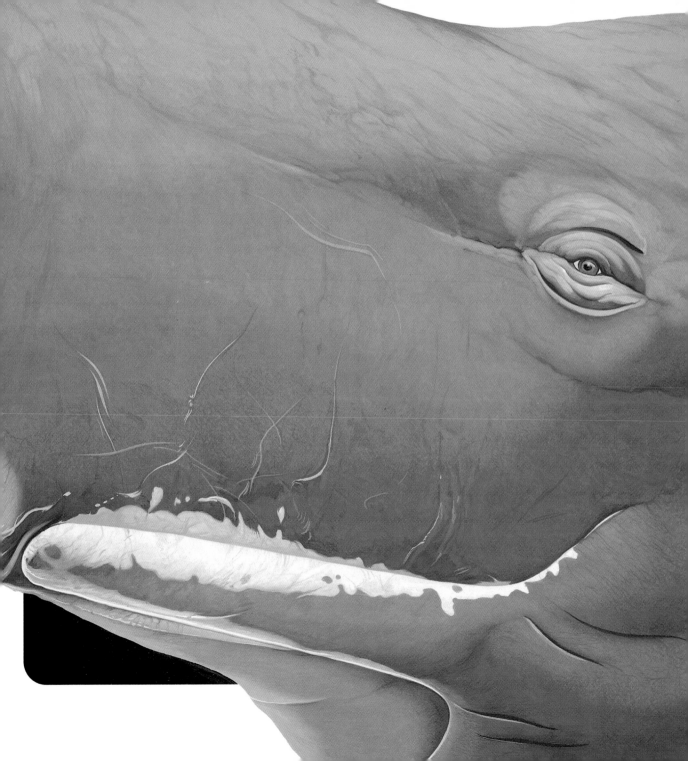

Facts Behind the Faces

The animals in this book live in different regions of the world—from northern coasts and tropical waters to the deepest parts of the oceans. This part of the book tells you more about these animals—where they live, their relatives, their eating habits, and their enemies. From the great sperm whale to the tiny deep-sea shrimp, here are the facts behind the faces.

Otter
Family: The weasel family includes otters, badgers, and skunks.
Other otters: Sea otters and African clawless otters.
Where they live: Rivers, lakes, and coasts on nearly every continent.
How they live: Often on their own, and mostly active at night.
What they eat: Frogs, fish, crabs, and crayfish.

Shore Crab
Family: The crustacean family includes crabs, lobsters, and crayfish.
Other crabs: Pea crabs, hermit crabs, spider crabs, and giant crabs.
Where they live: As far north as Norway to as far south as northern Africa
What they eat: Worms, mussels, smaller crabs, and other small sea creatures
Enemies: Many, including large fish and seagulls.

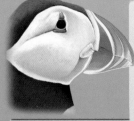

Puffin
Family: The auk family includes puffins, razorbills, guillemots, and auklets
Other puffins: Horned and tufted puffins, which live on the Pacific coast.
Where they live: Atlantic and Arctic coasts of North America and Europe.
How they live: Winters spent at sea, while spring and summer on land.
What they eat: Fish and shellfish.

Common Seal
Family: The seal family includes eared and earless seals and walruses.
Other seals: Earless seals include elephant, harp, and hooded seals.
Where they live: The coastal waters of the northern Atlantic and Pacific.
What they eat: Fish, shellfish, and squid.
Enemies: Many, including killer whales, sharks, and human beings.

Fulmar
Family: The petrel family includes fulmars, storm petrels, and shearwaters.
Where they live: Northern coasts of Canada, Greenland, and Europe.
How they live: Nesting on cliffs of coasts and islands.
What they eat: Fish and scraps from harbors and fishing boats.
Size: 20 inch body and 44 inch wingspan.

White-tailed Sea Eagle
Family: There are about 60 kinds of eagles in the world.
Other sea eagles: White-bellied and Steller's sea eagles.
Where they live: Iceland, Greenland, northern Europe, and Asia.
What they eat: Fish, seabirds, small mammals, and dead animals.
Size: Their wingspan can reach up to 8 feet.

Stingray
Family: The ray family includes skates, which are also diamond-shaped.
Other rays: Electric rays, eagle rays, and manta rays.
Where they live: All the world's oceans and South American rivers.
What they eat: Fish and shellfish.
Special features: A sharp spine on the tail gives a poisonous sting.

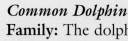

Common Dolphin
Family: The dolphin family is related to whales and porpoises.
Other dolphins: Bottle-nosed and humpbacked dolphins, killer whales.
Where they live: Across the world, mostly in warm and tropical waters.
What they eat: Fish and squid.
Enemies: Human beings—dolphins are hunted and caught in nets.

Cod

Family: The cod family includes haddock, whiting, saithe, and ling.
Where they live: The coastal waters of the northern Atlantic Ocean.
How they live: In large groups, often near the water's surface.
What they eat: Fish, worms, and shellfish.
Enemies: Human beings—people eat vast quantities of cod.

Deep-sea Shrimp

Family: The shrimp family is related to crabs, lobsters, and crayfish.
Other name: Some kinds of large shrimp are called prawns.
Where they live: In the deep waters of the open ocean.
What they eat: Small plants and animals and discarded food.
Enemies: Almost all sea creatures large enough to eat them.

Anglerfish

Family: There are over 200 types of anglerfish.
Related fish: Longlure frogfish, shortnose batfish, and football-fish.
Where they live: Coastlines from Scandinavia to northern Africa.
How they live: In winter they stay on the sea bed, close to the coast. In spring and summer they travel into deeper waters.

Fang-tooth Fish

Family: The fang-tooth is the only fish in its family.
Similar fish: Gibberfish and deep-sea prickle fish.
Where they live: In the dark, lower reaches of the open ocean.
What they eat: Other fish and small sea creatures.
Size: Around 5 inches.

Sperm Whale

Family: The whale family includes beaked whales and belugas.
Where they live: In oceans across the world.
What they eat: Fish, lobsters, and giant squid.
Enemies: Human beings, although most countries ban whale hunting.
Size: 36-66 feet in length.

Index